Planet Earth

Protecting the Oceans

EMILY HUTCHINSON

Globe Fearon Educational Publisher
Paramus, New Jersey

Paramount Publishing

Planet Earth Series:

Alternative Sources of Energy
Endangered Species
Global Warming
Our Disappearing Rain Forests
Protecting the Oceans

Editor: Erika Capin
Production editor: Teresa A. Holden
Cover and text design: London Road Design
Production: Execustaff

Cover photo: © Gary Bell/The Wildlife Collection
Woodblock illustration: London Road Design

Science Consultant: Richard Peterson
Richard Peterson holds a B.A. in Science from Reed
College in Portland, Oregon, and an M.S. in Education
from Lewis and Clark College in Portland, Oregon. He is
currently teaching biology and chemistry at Beaverton
High School in Beaverton, Oregon. Environmental
science and ecology are important components of
his classroom.

Library of Congress Catalog Card Number: 93–71615

ISBN   0–8224–3228–5

Printed in the United States of America

4. 10 9 8 7 6 5 4 3 2 1
MA

Contents

1 The Oceans and Our Environment 1

2 Protecting Life in the Oceans 15

3 The Problem of Overfishing 25

4 Oil Pollution in the Oceans 35

5 Dumping of Wastes in the Oceans 47

6 Activities of Greenpeace and Other

Organizations 57

7 What Can You Do? 67

Glossary 75

1

The Oceans and Our Environment

Thor Heyerdahl is famous for his voyages on primitive sailing vessels. In 1947 he sailed the Kon-Tiki, a balsa-log raft, from South America to Polynesia. It proved that early people could also have ridden the currents of the Pacific Ocean on such simple rafts. His later voyages proved that such travel was possible on the Atlantic and Indian Oceans as well. In the late 1960s, he built Ra I and Ra II, **papyrus** reed boats, to sail across the Atlantic. These boats were similar to ancient Egyptian boats. Heyerdahl's 1977–78 voyage was

Note: Words in **bold** type can be found in the glossary.

on another papyrus reed boat, the Tigris. But Thor Heyerdahl proved one thing that he hadn't expected. The oceans had become terribly polluted. Many parts of the Atlantic were dirty with floating globs of oil. Some of these globs were as large as a slice of bread. Plastic bottles floated on the oil-spotted water. It looked like a dirty port near a big city. He saw an Indian Ocean that looked like a soup of black tar. In this soup were bobbing cans, bottles, and other trash. There were logs, planks, boards, and large sheets of plywood. All the wood was smeared with the oil that polluted the water. No one on the crew of the Tigris could find a place even to dip a toothbrush. Heyerdahl burned the Tigris in 1978. That day, he sent a letter to the United Nations Secretary General. He said the burning was a protest against "the inhuman elements" of our world. He said that we were turning our world into a sinking ship.

Facts and Figures

- More than 70 percent of the Earth's surface is covered by oceans.
- Of all the water on Earth, 97 percent is in the oceans and seas.
- Life on Earth first appeared in the oceans.
- More than 80 percent of all living matter is in the oceans.
- The oceans play a major role in controlling our weather.

- If the Earth were as smooth as a marble, it would be covered with water to a depth of over 1-1/2 miles.

The World Ocean

When astronauts first blasted off from Earth and looked back at their home, they were surprised at how blue the planet is. That blue color, of course, comes from the water in the oceans. The oceans cover more than two-thirds of the surface of the Earth. In fact, some people say that the planet could very well have been named Ocean rather than Earth.

We usually think of the oceans as being separate. In fact, we have given each area its own name, even though the ocean is really one continuous body of water. We call the areas of this body of water the Pacific Ocean, the Atlantic Ocean, the Indian Ocean, and the Arctic Ocean.

The Pacific Ocean is by far the largest. It covers almost as much of the Earth's surface as all the other oceans put together. The Pacific is also the deepest ocean, with one section being over seven miles deep. If Mount Everest were placed in that part of the ocean, its top would still be beneath a mile and a half of water. The Pacific Ocean lies west of the Americas and east of Asia and Australia.

In order of size, the other oceans are the Atlantic, the Indian, and the Arctic. The Atlantic Ocean separates the Americas from Europe and Africa. It is about half the size of the Pacific

Ocean. The Indian Ocean lies in the area bounded by Africa, Australia, and Asia. The Indian Ocean is smaller than the Atlantic, but its average depth is greater. The Arctic Ocean lies north of Europe, North America, and Asia, and it is the shallowest and coldest of the oceans.

The words *ocean* and *sea* are often used to mean the same thing, but the seas are a little different from the oceans. Seas are smaller than oceans. Some seas, such as the Sargasso Sea in the Atlantic Ocean, are part of the open ocean. Other seas, such as the Mediterranean or the South China Sea, are partly enclosed by land. The water in seas is different from that in the oceans, too. It is either fresher or saltier, warmer or colder than the oceans.

Saltiness of the Ocean

The amount of salt dissolved (blended with water) in the ocean is called its **salinity.** Salt is present in the ocean in the amount of about 35 parts of salt to every thousand parts of water. This is an average amount. Some areas of the ocean are saltier than others. The salinity of the ocean has remained pretty much the same for millions of years. This is surprising, considering the fact that new salt is constantly being added to the ocean. Here's how it works.

As ocean water **evaporates** on the surface, the salt remains behind. The water vapor rises, forms clouds, and falls again in the form of rain. The fresh water from the rain and from rivers washes over the soil and rocks of the land, picking up salts and minerals as it goes. It eventually returns to the ocean, bringing those salts and minerals along. It would seem that the ocean would get saltier as time went by, but this is not the case. The salt becomes trapped with the mud and sand that build up on the floor of the ocean. Thus, the salt content of the water has remained just about the same for millions of years. If it hadn't, the many plants and animals that are used to a certain amount of salt would have died out.

Although the average amount of salt has remained the same, salinity in the ocean varies from place to place. In areas where there is a lot of rain or melting ice, the ocean is less salty. In areas where more water evaporates than falls as rain, the water is saltier. There is enough salt in the ocean to cover all the land on Earth with a layer of salt 520 feet thick.

Ocean Currents: Rivers in the Ocean

Just as the air travels on currents of wind, so do the waters of the ocean travel on currents. Warm water from the **tropics** moves toward the **polar regions.** This forces the cold water to move toward the tropics. This constant flowing of water helps to spread heat throughout the world.

These ocean currents have been called the "rivers in the ocean." Ships sailing east across the Atlantic Ocean can use the Gulf Stream and the North Atlantic Current. These currents help to speed the ships.

Winds help keep currents on the surface moving. Another factor that keeps currents moving is the spin of the Earth. Continents that block a current also have some effect on its flow. The current is forced to change direction when it comes up against a large land mass. The giant loops in the currents that are caused by wind, the Earth's spin, and land barriers are called **gyres.**

There are other kinds of currents besides those found near the surface. Deep-sea currents, called **drifts,** are not affected by the winds. Cold temperatures cause these drifts to keep moving. Cold water is heavier than warm water, so it sinks, just as cold air does when it sinks and takes the place of warm air. The cold water in the polar regions sinks toward the bottom of the ocean. It moves along the ocean floor toward the equator. Then it wells up to replace the surface water that is being blown offshore by the winds. As these cold currents travel through the tropics, they become warmed by the sun.

When deep, cold water wells up toward the surface, it brings with it **nutrients** from the ocean floor. These nutrients cause rapid growth of **phytoplankton,** tiny plants that are food for millions of ocean animals. Such areas of the ocean attract large numbers of both fish and the people who catch them.

Pollution and the Ocean

The constant movement of the ocean is one reason that water pollution is such a serious issue today. **Pollution** of waters in

Alaska, for example, could ultimately affect the waters of any other part of the world. Currents would eventually carry the problem elsewhere. The best proof of this is the fact that DDT, a **pesticide** used in farming, has been found in the fat of Antarctic penguins. Thousands of miles from its source, the DDT had been carried by ocean currents to the Antarctic.

One of the first people to recognize the problem of ocean pollution was Rachel Carson. She was very interested in the plant and animal life in the oceans. In 1961, she wrote a book called *The Sea Around Us.* Her next book, *Silent Spring,* was published in 1962. In it, she discussed the problem of pollution long before most people had ever thought about it. After the publication of this book, many people began to see that all aspects of nature are connected. She was the first to point out that the DDT and other pesticides that are used to control insects on a farm will eventually get to the ocean. There, they will affect marine life and poison the fish that people will eat.

No one is sure what would happen if ocean water became so polluted that it couldn't recover. Some areas of the ocean may already be so badly polluted that they can never be healthy again. Think about this: clouds are formed from water vapor that comes from the ocean. Pollution in that water vapor could affect our air and fresh water supplies as well. Because plant life on land depends on rain, all our food supplies would be affected, not just those foods that come from the ocean. The importance of keeping our oceans clean and healthy cannot be overstated.

Plankton: Food for the Creatures of the Oceans

Plankton are very tiny plants and animals that live in the ocean. They drift about on the currents and tides. Plant plankton is called phytoplankton, and animal plankton is called **zooplankton.** Both kinds of plankton provide food for nearly all **aquatic** animals.

Phytoplankton: Grass of the Sea   Phytoplankton are single-celled plants. Their name comes from the Greek words *phyto*, which means "plant," and *plankton*, which means "wandering" or "drifting." Ultimately, all life in the ocean depends on phytoplankton, for they are the first link in the food chain. A food chain is a group of organisms that depend on one another for food. In the oceans and seas, phytoplankton are eaten by small fish, the small fish are eaten by bigger fish, and the bigger fish are caught and eaten by humans. On land, the food chain often begins with grass. This is why phytoplankton are sometimes called "the grass of the sea."

It is possible that without phytoplankton, all life on the land, too, would be in serious trouble. That is because phytoplankton play such an important role in keeping our air clean. Each year these plants take out of the atmosphere about half of the **carbon dioxide** that we put in it. If this weren't done, our planet would be warmed quickly through the greenhouse effect. That is, heat would be trapped inside the atmosphere by a thick cloud of carbon dioxide. A warmer planet would cause some ice at the polar caps to melt. This would cause sea levels to rise, and there would be flooding in coastal areas. Of course, phytoplankton are not the only things that affect temperatures, but they play an important role.

The Needs of Phytoplankton   Phytoplankton have certain basic needs. These tiny plants all need nitrogen and phosphorus as nutrients. They also need sunlight for energy. Different species (types) of phytoplankton have other needs. For example, some species will be killed if the water had just a little too much nickel (a type of metal) in it.

Some unknown factors seem to affect the growth of phytoplankton. In many parts of the ocean right now, for example, conditions seem perfect for phytoplankton. There is enough nitrogen and phosphorus to support large amounts of the tiny plant. But for some reason, there are far less phytoplankton than should be expected. No one is quite sure why this is so.

Phytoplankton and Ultraviolet Light   Ultraviolet light from the sun is harmful to many types of phytoplankton. The ozone layer in the atmosphere has, until recently, protected our planet from harmful levels of ultra-violet light. But now there is a hole in that protective layer. The hole is largest over Antarctica, and some scientists are very concerned that phytoplankton will be affected.

Other scientists say the problem is not that serious. The larger species of phytoplankton are not seriously affected by the ultraviolet rays. There is no real danger to the food chain, some say. Others say that the smaller phytoplankton might be important in ways that we don't know about.

Pollution and Phytoplankton   Most kinds of phytoplankton grow along shallow coastal areas, around **reefs,** and at **upwellings** of deep water. An upwelling is a place in the ocean where cold water from below moves up. This moving water brings nutrients to the warmer upper layers. It provides fertilizer for the phytoplankton near the surface. So this is where the marine life that feeds on phytoplankton is found. Unfortunately, these areas tend to be affected the most by pollution. Over 80 percent of all ocean pollution comes from activities on land. This pollution takes the form of oil, sewage, heavy metals, and certain pesticides.

Zooplankton   In addition to the phytoplankton, tiny animals called zooplankton also live in the ocean. Zooplankton are often larger than phytoplankton, and some species can be seen with the naked eye. One example is **krill,** an animal that looks something like a shrimp but is about 2.4 inches long. Whales eat them by the ton. Krill survive by eating the phytoplankton.

One of the greatest mysteries about zooplankton is why so many species of them **migrate** up and down daily. Many of them move up toward the surface each evening and then go back down deeper at dawn. One species, which is less than an eighth of an inch long, moves up and down almost 1,300 feet each day.

Oceans and the Weather

The oceans not only affect the weather, but they are also affected by it. Wind, sun, and rain each play a role in creating ocean currents. And the currents either warm or cool the climates of the areas they touch.

Land absorbs and loses heat quickly, unlike water. Think about the moon for a moment. It is about the same distance from the sun as the Earth is. Yet the side of the moon that faces the sun is extremely hot, and the side that does not face the sun is extremely cold. The Earth would have similar extreme temperatures if it didn't have the oceans and the atmosphere. The atmosphere traps the planet's warmth and keeps it from escaping into space. Without the oceans and the atmosphere, the land would be either too hot or too cold, depending on whether it was facing the sun.

The oceans absorb and lose heat more slowly than the land, and the currents help to carry the heat around the whole planet. The surface of the ocean is warmed by the sun. Because warm water is lighter in weight than cold water, a layer of warm water stays near the surface. The winds blow this top layer in a certain direction, and the colder water below moves up to take its place. Then the colder water is warmed, and the cycle continues.

When the warm, moving water reaches a coast, it warms the land. The warm currents help to keep coastal areas warmer than inland areas in the winter. This is why winters in London are not as harsh as winters in New York, even though New York is farther south than London. London benefits from the warmth of the Gulf Stream waters.

In some areas, where cold water wells up to replace the warm water, it cools the land. The cold currents from the deeper parts of the ocean carry the mineral salts from the deep. This creates an **environment** that encourages the growth of phytoplankton, which in turn provides food for the creatures of the sea.

Ocean Fronts

The ocean, like the land, has areas where there is little life. Such areas can be compared to the deserts on land. The ocean also has areas where life is abundant. Such areas can be compared to the rain forests on land. These areas include coastal wetlands, **estuaries,** and reefs. The Great Barrier Reef of northeastern Australia, for example, is home to more than 3,000 animal species.

Surprisingly, one area that has a great deal of life is right around the continent of Antarctica. This part of the ocean is called the Antarctic Convergence. Here, currents of cold water traveling north meet with currents of warm water from the tropics traveling south. This creates a turbulence (disturbance) that stirs up the bottom. Water from the bottom, rich with nutrients, comes to the surface. These nutrients provide food for penguins, seals, squid, and whales.

Areas where cold and warm water meet are called **ocean fronts.** The surface temperature across fronts may be warmer or cooler than the surrounding water by 15 to 20 degrees Fahrenheit. Sometimes these temperature differences create fog banks over the cooler water.

A front is a kind of oasis in the middle of the sea, as far as the marine animals are concerned. The stirring up of the waters here brings nutrients from below, providing food. Schools of fish often follow these fronts. This, in turn, brings feeding seabirds to the area. In the past, fishermen would locate fronts by looking for the feeding seabirds. Now, fishermen can use information provided by satellites about fronts. Thermal (heat-sensing) images from satellites can reveal areas where fronts occur.

Unfortunately, these fronts provide more than just nutrients for sea animals. The fronts are areas where pollution caused by human beings collects. Small plastic pellets bob on the surface, and seabirds mistake them for food. These pellets block the birds' digestive tracts, causing weakness and eventually starvation. Sea turtles eat sheets of plastic and six-pack rings because they look like jellyfish. These items block the digestive tracts of the turtles. In addition, oil spills that become part of an ocean front don't break down as easily as they do in the open ocean. Seabirds who feed at the ocean surface can become coated with oil. Their feathers lose their ability to repel (hold back) water. The birds swallow oil as they try to clean themselves.

Pollution that is washed from rivers is not distributed evenly in the ocean. At ocean fronts, it may be 1,000 to 10,000 times more concentrated. Any animal that feeds at such ocean fronts will, of course, be affected by these pollutants. Once pollutants get into the food chain, it is likely they will end up in food eaten by people.

2

Protecting Life
in the Oceans

Philippe and Caron looked through the windows
of their face masks. Bubbles rose in the water
around them. They breathed evenly, following the
directions they had received in their diving class.
They saw a wall of coral, its gold, red, and yellow
colors sparkling in the blue water. Red-speckled
fish, blue-striped fish, and yellow-dotted fish swam
in and out of the holes in the coral. They saw the
open shells of giant clams revealing emerald green
and blue flesh inside. Sponges held on to the wall
of coral, as if afraid they'd fall off. The beauty of

this underwater world seemed to cast a spell over the two divers. As Philippe reached out to break off a piece of the coral, Caron stopped him. She gestured a "No" with her finger, shaking her head at the same time. Philippe had almost forgotten. For a moment, he had thought a souvenir piece of coral would look great in his living room at home. He was glad Caron had reminded him of something he already knew. Coral is a living **organism.** *Breaking off a piece of it affects other parts of the coral nearby. A small break can kill a large section. Philippe would just have to take a picture with his underwater camera so he could always remember this scene.*

Just as on land, life in the oceans is not evenly distributed. Some areas are rich with a variety of living things, and some are very poor. The areas that are richest in ocean life are in coastal zones. This is where nutrients wash down from the land. It is also where winds and ocean currents help to dredge up nutrients from the ocean floor. The mixing of salt water and fresh water also adds to the wide variety of life in these areas. This is because there are both saltwater and freshwater plants and animals living together.

There are four main **ecosystems** that support ocean life. An ecosystem is a community of various kinds of plant and animal life, including their surrounding environment. These ecosystems are **saltmarshes, mangroves,** estuaries, and **coral reefs.** Saltmarshes and mangroves are quite similar. The main difference is their location. Saltmarshes are tidal wetlands in temperate zones. Mangroves are tidal wetlands in tropical zones. In both areas, the main plants that grow offshore are seagrasses. These are true flowering plants that bloom under the water. In the temperate zones, they provide food for ducks and geese. In the tropics, they provide food for sea turtles and various sea mammals. In both zones, the plant life attracts many **finfish** and such **shellfish** as shrimp. The plants also filter out pollution, slow down the waves and currents, and prevent the erosion, or wearing away, of the coastline.

Estuaries are places where rivers and oceans meet. These areas are extremely fertile, and they support a large variety of ocean life. Crabs, oysters, mussels, and prawns are found here. This is also where many ocean fish start their lives before they go out into deeper waters. Of all the fish caught in the continental shelf of the eastern United States, at least 75 percent spend part of their lives in estuaries.

The ecosystems that support the greatest variety of life are the tropical coral reefs. They support one-third of all the species of fish. They appear to be the world's oldest ecosystems, having survived since life first began. They have more plant and animal species than any other ecosystem. Some of these plants and animals supply human beings with valuable medicines. Some antihistamines, hormones, and antibiotics are made from materials found in coral reefs.

Protecting Coral Reefs

Tiny animals called **polyps** build huge coral reefs. They do this by using chemicals from the ocean water to build hard skeletons around their soft bodies. Tiny plants living inside the polyps help in the process. Millions of polyps live together in huge colonies. When they die, layers of their skeletons are left, forming the coral. Coral grows about as fast as a human's fingernails.

The Great Barrier Reef has been growing for at least 15 million years. It is the biggest structure ever built by any living creature or group of creatures. It stretches for 1,260 miles off the northeast coast of Queensland, Australia. This reef is so big that it can be seen from the moon. More than 350 species of coral grow here. The coral is home to more than 1,400 species of fish, uncountable sponges, and other sea creatures such as sea urchins and brittle stars.

This beautiful reef, sadly, is in danger. In addition to the problems created by humans, one of the inhabitants of the reef is doing damage to its home. The crown-of-thorns starfish, which can grow as large as 16 inches across, began appearing in large numbers in the 1960s. By the 1970s, the starfish had stripped large areas of the reef of its living coral. Bare limestone was left behind. A single starfish may eat up to 16 square inches of coral in a day. Since the 1970s, the numbers of these starfish have been going down. The coral has been able to start growing back, although it may take 30 or 40 years to get back to where it was. Scientists who study the reef have found evidence that similar problems occurred before. Some think that such "problems" may simply be a part of the normal cycle of life in the reef. In 1980, Barrier Reef Marine Park was set up to protect part of this area. Australia set aside about 5 percent of the reef as a preserve.

How Coral Reefs Are Damaged   Most coral reefs grow in tropical waters where temperatures stay between 77 and 84 degrees Fahrenheit. They can grow only in clean, unmuddied water. The amount of light they need is found only in shallow waters (no deeper than 100 feet), usually close to land. This makes them easy for people to use. People can make a lot of money by destroying coral reefs. The coral makes good building material when it is cut into slabs. When it is ground up into powder, it makes good polishing material. When it is burned, the result is lime, a material used to make mortar, cement, and fertilizer. But using the raw materials of a coral reef for such things as these is not very wise. It is something like tearing up your best clothes to provide pieces of cloth for washing the car.

Another way coral reefs are damaged is through blast fishing. This method involves setting off explosions near the coral beds. The power of the blast stuns the fish in the area, which are then easier to catch. It also stirs up sediment (matter that sinks to the ocean floor). This creates an environment in which coral can't grow.

The Importance of Coral Reefs   Coral reefs are most valuable when they are living. When living coral is cut, it harms the remaining coral as well. The coral reefs of Bali, an island in Indonesia, have been damaged so severely that the coast is now being worn away by the waves. Without the coral to act as a buffer, the waves hit the land with full force. In Sri Lanka, the government is trying to save the coral reefs that have already been damaged. They have banned reef mining to prevent more damage. Many people have lost their jobs because of this. In one area, the government is paying 10,000 people to make up for their loss of income due to the ban.

It is often difficult to get countries to agree to preserve their own natural resources if it means they'll have less money. When the tourist industry is involved, however, it is easier. This is true especially if the tourists are coming to see the beauty of an area. One good example of tourism helping to preserve a place is Bonaire, a Caribbean island. The biggest attraction of this island for tourists is its large coral reef. This, along with the mangrove forests and seagrass beds along the coast, also supports the local fishing industry. Natives of the island harvest a variety of reef fishes, conch, spiny lobster, and turtles. Their catches are limited by local law, partly because the tourist industry is so large.

Visitors to Bonaire want to enjoy deep-sea diving and underwater photography. So it is in the best interests of the islanders to preserve the underwater beauty. Inexperienced divers often damage coral reefs by breaking the coral accidentally. Yachting and boating also can damage the reef. The Bonaire Marine Park, established in 1979, protects the area. This is one example of an area where tourism and scientific research work well together. **Marine biologists** (scientists who study life in the ocean) help to manage the area. The money from tourism allows the islanders to conserve their resources. And the scientists help identify areas that need extra protection.

It was a still summer evening on the western Atlantic Ocean near the New Brunswick shore. Erika and Cesar stood on the deck of the ship and looked out. No planes were in the sky. No other boats could be seen. It was so quiet that it almost seemed as though they were alone in the universe. Suddenly they heard a sharp sound. It sounded like an explosion. A few minutes passed, and the sound was repeated. Worried, Erika and Cesar hurried below to talk to Ishmael, their friend. When Ishmael came up on deck and heard the sound, he laughed. "I assure you, that is no explosion," he said. Then Ishmael told Erika and Cesar what the sound was. He told them to look through their binoculars. "See that baby whale? It's a right whale. Because it's about 20 feet long, it must be about 6 months old. Now watch for a few minutes. You'll see its mother." Sure enough, a few minutes later, the mother surfaced near the baby. Happily, the calf hurried over and slid up on her back. But she batted him away with a flipper. She might have been playing, but it is more likely that she was annoyed. The calf had interrupted her while she was trying to get some dinner. Like babies everywhere, the baby right whale missed its mother when she was gone.

Whales in Danger

Whales are among the animals that have nearly been wiped out by humans. There are several species of whales. Some species are in more danger than others. Whales have been hunted for several reasons: for their oil, meat, blubber, and **ambergris.** Ambergris is a substance found in the stomachs of some whales. It is used to make some perfumes.

In the early days of whaling, it seemed that the supply was endless. One species, called the "right" whale because it was the right one to hunt, was slow and easy to catch. Hunted from small boats with hand lances, these whales soon showed signs of being overhunted. They were no longer so easy to find.

So the whalers changed their prey and their methods. Forced to hunt for faster whales, such as the blue, fin, and sei whales, they needed faster boats and stronger harpoons. Better methods of whaling were developed. Once again, the stocks were overhunted. Whalers soon turned to new areas, such as the Antarctic. Now it appears that whales may soon become extinct (die out completely).

There are only a few thousand blue whales left out of the 100,000 we think existed at one time. These might not be enough for the population to recover. Some scientists think that it is too hard for the whales to find each other to reproduce. Even sperm whales, which used to be caught at a rate of 30,000 a year, are now seen only in groups of 10 or less.

Sperm whales are the most numerous of the great whales. They are also the most valuable to people. The fatty material in their foreheads produces a higher-quality oil than is found in other whales. This oil performs well even under conditions of extreme heat, cold, or pressure. Luckily for the sperm whale, a substitute has been found. Jojoba oil, found in a desert plant, works just as well if not better. It is used in candles, shampoos, face creams, waxes, and industrial lubricants. The only problem is that the jojoba shrub grows very slowly in the wild. It is difficult to cultivate (grow on a farm). So the slaughter of the sperm whale for its oil will probably continue until a way can be found to get enough jojoba oil.

Recently there has been a growing awareness worldwide about whales. People concerned about the environment started to use the whale as a symbol. Perhaps because the whale is so big, so dramatic, and so mysterious, it became a powerful symbol. This awareness grew into the "Save the Whales" movement. Petitions were signed to stop the whaling industry from destroying these creatures.

In spite of the efforts of the anti-whaling movement, whaling continues. The International Whaling Convention (IWC) declared a moratorium (ban) on most whaling. This ban was supposed to start in 1985. However, many countries do not observe it. Iceland, Norway, and Japan say they are catching whales for "scientific purposes." Several other countries simply ignore the ban. Scientists say that whales need protection. People must stop hunting them, if the whale population is ever to be restored to its former numbers.

The California Sea Otter

The California sea otter lives among the kelp beds off the coast of California. Kelp is a large, brown algae that looks something like seaweed. This animal is the only sea mammal that doesn't have a layer of blubber to keep it warm. Because of this, it must spend a lot of time eating to keep warm. This animal even feeds

at night. Each sea otter eats over two tons of seafood a year. It eats fish, urchins, abalones, crabs, snails, clams, and sea cucumbers. Because of its huge appetite, it has made enemies among people who sell shellfish.

The California sea otter was heavily hunted for its fur for over 150 years and it was almost extinct by 1911. In 1937, some survivors were found and protected, and their numbers have been increasing. Today, there are about 1,700 of them on 160 miles of California coastline.

Sea otters feed on sea urchins, which feed on kelp and other seaweeds. Where there are no sea otters, sea urchins are plentiful and there is very little kelp. Where there are sea otters, the kelp harvesters do not have to worry about the sea urchins. The arguments about the value of the sea otters continue. Kelp harvesters benefit from the presence of the sea otters. Kelp beds support a great wealth of fish as well, so the fishing industry benefits, too. But shellfish harvesters do not like the sea otters, who eat their catch. The argument about whether this animal is a pest or a help is still going on.

Meanwhile, the sea otters continue to become trapped and drowned in fishing nets. They are also very easily harmed by any oil in the water. Once they get oil on their fur, they lose any insulating (heat-keeping) ability and freeze to death.

3
The Problem of Overfishing

It was Jeremy's first day at sea as a fisherman, and he was really enjoying it. He loved the wind in his face. He loved the smell of the salty air. And he loved the sounds of the birds flying overhead. Today, the fishermen were looking for pollack, a fish that looks something like cod but has a darker color. Jeremy knew that the captain had a big contract for the **roe,** the eggs of the pollack. The crew had put the nets out a few hours before. It was time to haul them in. Jeremy couldn't help but feel excited when he saw how many pollack were *caught in the nets. He helped get the eggs from the females by cutting them out. In just a few hours, they had more pounds of roe than Jeremy would have thought possible. He was looking forward to the next step in processing the fish. He wanted to see how fish were frozen and stored at sea. To his complete surprise, he saw his fellow fishermen start throwing the dead male fish overboard. "What are you doing?" he cried. "That's against the law, and it has been for years!"*

"If you want to keep your job, Jeremy, you'll do what you're told," said the captain. "We need room for more roe. No one will ever know, as long as we freeze enough fish to account for the roe. We've always done it this way. We're not going to change now." Jeremy was shocked. Suddenly it didn't seem like such a beautiful day or such a wonderful job.

Not so long ago, the oceans were seen as an almost limitless source of food. Modern methods of fishing, however, have done extreme damage to this food source. These modern methods include the use of airplanes to spot schools of fish in surface waters. They include the use of satellite readings of water temperatures. They also include echo sounders and **sonar** to look for fish in deep waters. In some cases, power-driven machines are used to raise and lower fishing nets. Bright lights are sometimes lowered into the ocean to attract fish. The fish are then sucked into the ship with powerful vacuum pumps. Another method of catching large numbers of fish is called **long-lining.** Long lines with scores of hooks on them are stretched out across the sea. The fish are caught on these hooks.

Huge floating factories go out to sea with the fishing boats. Such a factory ship can process the fish from up to 30 fishing boats at a time. They have machines that will clean, freeze, and can the catch on the spot.

Fewer Fish

These and other methods have created a problem called **overfishing.** This means that too many fish are being caught. The fish can't reproduce fast enough to replace the previous year's catch. Scientists say that 14 of the most valuable fish are in danger of "commercial extinction." Soon there won't be enough of these species left to make it worthwhile to fish for them. Included in this group are red snapper, swordfish, striped bass, and Atlantic bluefin tuna. One way to save these fish is to put a ban on catching them for several years. But such a ban would put a lot of fishing companies out of business.

Solving the Problem

There are other ways to control the problems of overfishing besides a ban. One of these ways is to gradually limit fishermen's days at sea. Another way is to force fishermen to use nets with bigger holes in them. In many cases, the nets with smaller mesh trap fish that are too small or otherwise undesirable. These fish, called **bycatch,** are then thrown back into the water, dead or dying. Such waste has been called "a national scandal."

Nets with small mesh are not the only cause of bycatch waste. Some fishermen throw away large numbers of caught fish to save room on their boats for more valuable species. In 1990, ship captains in Alaska reported throwing away 25 million pounds of halibut and 550 million pounds of groundfish for this reason. The actual figures are probably much higher. It has been estimated that for every pound of shrimp caught, 9 pounds of other fish caught in the same net are tossed overboard. Unfortunately, these fish are not tossed quickly enough for them to survive.

Traditionally, fishing has been an industry that anyone with a boat can enter. To control competition for the most valuable fish, "seasons" are sometimes declared, as in hunting. The September 1991 season for Pacific halibut in Alaskan waters is a good example of the problem. This fish was once harvested in a 6-month season. Now the season lasts only 2 days. In those 2 days, 23.7 million pounds were caught, but about 4 million pounds rotted. That was because they couldn't be frozen fast enough.

One solution that has been suggested is to have a system of individual transferable quotas (ITQs). This means that a boat owner would get a permit for a certain amount of finfish or shellfish each year. These permits could be leased, sold, or passed on in a family. Some fishermen don't like this idea. They think that corporations will buy up the permits. This would lead to the small, independent fishing companies being swallowed up by big companies.

In 1976, the Fisheries Conservation Management Act was passed by Congress. This act extended the United States area of control (and exclusive fishing rights) from 12 miles to 200 miles offshore. This act kept foreign fishing boats out of the United States coastal waters, but it did nothing to keep Americans from overfishing. Some coastal areas have established Exclusive Economic Zones (EEZs) to help solve the problem. These zones extend 200 miles out to sea, and they are controlled by law. Quotas are established for numbers of fish caught. Within these EEZs, important fish habitats are protected.

Fishermen who object to quotas, permits, and the like will soon have to face facts. The remaining stocks of fish in the oceans will have to be carefully managed if they are to survive. Overfishing, if allowed to continue, will mean the dying out of one species after another. If that happens, there will soon be no fish left to catch.

Drift Nets: Walls of Death

In the early 1980s, a new type of net came into use. Fishing fleets from Japan, Taiwan, and South Korea began to use drift nets. These nets are made of thin, very strong nylon, and they are almost invisible. The nets are huge—as big as 50 feet deep and 30 miles long. They are suspended below the surface of the water with buoys. Weights at the bottom hold them vertically in the water, like drapes on a window.

One ship might use several dozen of these nets, lowering them in the evening. It is estimated that 2,500 ships put out more than 50,000 miles of drift nets every night. Drifting all night, the nets trap any animal that comes along. Fish become caught in the mesh of the nets, and their gills get tangled. They slowly die from suffocation. In the morning, the nets are hauled in, with their huge catches. Sometimes only 50 percent of the catch is usable. Parts of the catch might include birds, sea mammals, and trash.

Gill nets are smaller and are used in waters close to the coast. They work in much the same way as drift nets, trapping anything that comes along. For example, a group of gill nets in the North Pacific were set to catch salmon. In one year, the gill nets caught the following animals: 750,000 seabirds, 20,000 Dall's porpoises, 700 fur seals, and many small whales.

In 1987, a group in Canada tested some drift nets. One ship alone killed more than 200 dolphins and other sea mammals, in addition to more than 1,000 seabirds. Right after that, Canada banned the use of drift nets in its waters. Other countries have followed Canada's lead. New Zealand, for example, won't even allow drift net boats to use its ports. Ships that only want to take on supplies are turned away if they use drift nets. The United Nations General Assembly has declared a moratorium on drift net fishing in international waters. Unfortunately, the United Nations cannot force countries to participate in this ban on drift nets.

One of the saddest facts about these drift nets is that some of them have broken free of their boats. Such nets drift through the ocean for years, because they do not fall apart. The damage they will continue to do cannot be measured. It is no wonder that they have come to be called "walls of death."

Dolphins, Tuna, and Purse-Seine Nets

Purse-seine nets are so big that it takes two boats to set them in the water. The fishermen place the large net around a school of fish and then bring the ends together, closing it like a purse. These nets are designed to catch tuna, but they also capture any dolphins, harp seals, turtles, and other animals that happen to be in the area. Over 2,000 sea otters were drowned in such nets off the coast of California in a 10-year period. The use of these nets is now forbidden in any area where California sea otters live.

In the eastern Pacific Ocean, from southern California to Chile, tuna fishermen actually look for dolphins. For unknown reasons, schools of yellowfin tuna travel with the dolphins in these waters and nowhere else. Dolphins are easy to see because they must stay near the surface to breathe. They travel in groups of 50 to 1,000. When the fishermen in these areas spot a group of dolphins, they know that the yellowfin tuna will often be found swimming below them. The fishermen use their huge purse-seine nets to catch the tuna, and in the process they trap the dolphins, too. Because the dolphins need to breathe air, they quickly drown in the nets. It is estimated that 100,000 dolphins die each year when they are trapped in tuna nets.

Recently, some companies in the business of canning tuna made a decision that was favorable to the dolphins. Many people had been threatening to stop buying tuna caught in nets that were harmful to dolphins. In 1990, the world's largest tuna canning company started putting a "Dolphin Safe" symbol on its cans. This symbol indicated that the company no longer was buying tuna caught in purse-seine nets. Other big tuna canners followed this company's example.

Of course, cans of dolphin safe tuna cost a little more because the fleets are catching less tuna. Fishing fleets might have to try to fool the tuna by using floating logs to imitate the dolphins. In the future, any tuna swimming under a dolphin might be in the safest possible place, at least if it wants to avoid being caught by humans.

Sea Turtles

There are seven species of sea turtles. All of them are threatened by humans. Their situation is even more difficult than that of the whales because they do not reproduce at sea. This means that they can be caught easily when they come to the beaches to lay eggs. When they nest on beaches, people and other predators take their eggs.

Grown turtles are also hunted at sea. One species is used to make green turtle soup, an expensive, luxury food. Another species, the hawksbill, provides tortoiseshell, a product used in making craft items. The skin of sea turtles is made into expensive leather.

People who try to protect the turtles start with protecting the beaches where they lay their eggs. The Mexican Departamento de Pesca (Department of Fishing) guards one nesting beach with armed marines during turtle breeding season. An agreement to stop trading in sea turtles and sea turtle products has been signed by 89 nations.

In Mexico, where the world's most endangered turtles get tangled up in the purse-seine nets, fishermen use a special device. It is called the Turtle Excluder Device or the Trawling Efficiency Device (TED). This device is a metal or nylon grid that is inserted into the neck of the net. The grid is small enough so that shrimp can pass through it and remain netted. Turtles and other larger animals, however, are stopped. Because the TED is placed at an angle, the turtle slides toward the side of the net. Then it can get out through an opening. The use of the TED has reduced turtle capture by 97 percent. It has also increased shrimp catches, since more room is left in the net for the shrimp.

4
Oil Pollution in the Oceans

I t was a beautiful, crisp March day in 1989.
Tanisha and Olaf were seeing Alaska for the first
time. It was too bad that it wasn't the trip they had
planned. They had wanted to go to Alaska to see
its spectacular beauty. Instead, they were part of a
volunteer clean-up team. Tanisha and Olaf wouldn't
have any time to gaze at glaciers. They were too
busy trying to get the oil off the feathers of a bird.
Other volunteers were trying to remove oil from
the fur of sea otters. A few days earlier, on
March 24, an Exxon oil tanker called the Valdez

had run aground near Valdez, Alaska, in Prince William Sound. Tanisha and Olaf were just two of the more than 11,000 workers who tried to clean up the mess. They were using high-pressure water jets to wash the bird's feathers. The bird was so frightened that it took two of them to do the job. Olaf held the bird still while Tanisha cleaned the feathers. After working on the bird for two hours, Tanisha said, "I think we've saved this one. Let's see what we can do for the next one."

Next to Tanisha and Olaf was a volunteer by the name of Shawna. She was trying to save a sea otter. She had to calm the animal down by giving it a sedative. Then she flushed its eyes with water. After that, she washed and dried its coat. Then, she gave it some medicine. The animal appeared to be suffering from stomach pains. Maybe it had swallowed some of the oil. Shawna said, "I hope this medicine helps. It would be wonderful to save all these animals, but I don't like our chances." It turned out that Shawna was right. Volunteers at Prince William Sound worked day and night. But they could save only a small percentage of the marine animals who were affected by the spill.

The oil spill caused by the *Exxon Valdez* in March 1989 was one of the worst disasters caused by humans. While being steered to avoid icebergs, the 987-foot oil tanker rammed into an underwater **shoal.** The ship split open, and more than 11 million gallons of oil began pouring into Prince William Sound. By the next day, the oil spill had slowed to a dribble. By then, the oil slick was 8 miles long and 4 miles wide. Winds and currents over the next few days spread the oil slick, causing damage to 1,200 miles of coastline. If the accident had taken place on the East Coast, it would have dirtied the coast from Massachusetts to North Carolina.

Cleanup Methods Used

Within ten hours of the spill, cleanup workers were on the beach. The first thing they tried was oil-containment booms. These are floating barriers that block the spreading of oil in water. Once the oil is blocked, it can be removed with pumps or skimmers. Luckily, the water in Prince William Sound was calm. Booms work only in calm waters. But there was too much oil for the number of booms the workers had. It was impossible to remove all the oil this way.

Workers used rakes, shovels, and paper towels to clean up the oil from the land. On the beach, they wiped the oil off one rock at a time. When the next tide came in, the rocks became covered with oil again. The next method they tried was water jets, but the pressure was harmful to mussels, snails, and other sea life.

Finally, they tried an experiment called **bioremediation.** This means that scientists used living organisms to clean the oil. By spraying a fertilizer mix on the shore, they hoped to cause tiny oil-eating bacteria to grow. These bacteria are naturally present in the environment. The fertilizer used by the scientists just encouraged many of the bacteria to grow. This experiment had some success, but it wasn't enough.

The people who worked to save the animals used a variety of methods. They used high-pressure water jets to clean animals' fur and feathers. They scrubbed oil from sea otters' fur and gave them medicine to treat poisoning. They treated the animals for hypothermia (extreme coldness). They also treated them for dehydration (not enough water) and other conditions.

Immediate Effects of the Exxon Valdez Spill

The timing for the fishermen in the area couldn't have been worse. The herring and pink salmon were just about to start their springtime runs. The waters in the area were also rich in whales, otters, seals, porpoises, dolphins, and many kinds of birds. Over the next few months, workers struggled to save as many animals as they could. Most of the animals that came in contact with the oil died.

At least 100,000 sea birds died after their feathers became clogged with oil. The birds were unable to stay warm and died from the cold. Other birds died after getting oil inside their bodies. These birds' intestines became coated with the oil. This kept them from absorbing water and nutrients. After the spill,

151 rare bald eagles were found dead. The eagles had gotten oil inside their bodies, probably by feeding on oil-covered prey. The birds that were saved by the workers are likely to develop liver and kidney problems later on.

In addition to these birds, workers counted over 1,000 sea otters that died. Scientists think that the total was more like 3,500 to 5,500. Also, hundreds of thousands of other animals, including fish, seals, and shellfish, died. In the summer after the spill, 348 otters were cleaned and given medical treatment. Of these, 226 lived.

Other Oil Spills

Each day, there are hundreds— maybe even thousands—of minor oil spills. These do not receive much attention in the newspapers. It is only the major oil spills that we hear about. But the size of an oil spill is sometimes not a good measure of the damage it can do. For example, there was a spill of 140 million gallons of oil in the Gulf of Mexico in 1979, when an oil well blew up. This oil was quickly moved by ocean currents and winds. It appears to have caused little environmental damage. Yet the much smaller spill that occurred in the English Channel in 1978 was far more serious. When the *Amoco Cadiz* ran aground off the coast of France, it spilled 69 million gallons of oil. This caused widespread damage to 150 miles of beaches and other habitats nearby. It is obvious that the location of a spill has a lot to do with how well the environment can handle it.

The main source of all oil pollution of the oceans is the shipping industry. The oil spills reported in the newspapers are only a small part of the problem. Tens of thousands of smaller spills happen each year. The exact number is not known because not all of them are reported. In the United States alone, about 13,000 oil spills of different sizes are reported each year.

One cause of oil pollution in the ocean used to be the washing out of storage tanks in ships. Before a ship came into a port to take on a new load of oil, its tanks were washed out. The oily water that was released caused globs of floating tar in the ocean. New, safer methods include having the ships keep any leftover oil in a special holding tank, to be disposed of later.

Another source of oil pollution of the ocean is the people who change their own car oil. The Environmental Protection Agency (EPA) says that home mechanics in the United States collect about 100 million gallons of used car oil each year. Only about 10 percent of this oil is disposed of properly. The rest of it is dumped into the environment—on the ground, in the gutter, or down a drain. This amount is equal to more than 8 *Exxon Valdez* spills. Eventually, oil dumped on the ground gets into rivers, which then empty into the ocean. Oil dumped down drains gets to sewage treatment plants and eventually into the ocean. The United States Coast Guard says that sewage treatment plants alone pump twice as much oil into the ocean as tanker accidents do.

Breakdowns of Oil Spills

The first oil well was drilled over 130 years ago. Since then, oil has been an important source of energy for human beings. Long before humans ever started using petroleum (oil) products, however, crude oil was seeping into the ocean.

Crude oil is a natural product formed from organic (living) materials, including plants and animals from the oceans. This organic material decayed (broke down). More layers of organic matter covered the first layers. After long periods of time, the bottom layers were deep underground. The deepest layers were under a lot of pressure from the layers above, and the temperature was very high. Over time, heat and pressure changed the organic matter into oil. Only a small percentage of this oil remains trapped below the ground. This is the oil that we pump out with oil wells. Most of the untrapped oil seeps gradually into the oceans.

As this oil seeps to the surface, it is recycled naturally. It is broken down by natural chemical processes. We know, therefore, that the ocean can handle a certain amount of oil. But huge amounts all at once, as in a spill from an oil tanker, are very harmful.

Here is what happens after an oil spill. First, as the oil is released into the water, it begins to spread. This is an important step in the eventual natural breakdown of the oil. As the oil spreads, its surface area increases, allowing greater exposure to air, sunlight, and the water below. Eventually, these elements will cause the breakdown of the oil. Unfortunately, as the oil slick spreads, it becomes more difficult for people to remove it.

Once the oil has spread, it can evaporate more quickly. The speed of evaporation is fastest during the early stages of the spill. Evaporation takes place from within a few days to a few weeks. After that, the oil that is left starts to turn into a thick sludge and tar balls. Sometimes it forms what is called a "mousse." This is a thick mixture of oil and water. This mousse can sink or be washed onto the shore. Sometimes tar lumps are formed. These take a long time to break down, and they are especially annoying when they are washed up on beaches. It takes many years for them to disappear.

If an oil slick spreads out far enough so that it is thin, the sun can break it down quickly. When people try to contain a slick, they keep it in a smaller area. This makes the slick deeper or thicker, which makes it harder to **biodegrade** (break down) naturally. To picture this, imagine that you have 1,000 ice cubes in a bathtub full of water. If you leave them alone, they will spread out in a thin layer and melt fairly quickly. If, however, you gather them all together in a small part of the tub, they will be in a thicker layer. Less of their surface will be exposed, so they will melt less quickly. This is something like what would happen with an oil slick contained in a small area. It would be easier for people to remove the oil by skimming, vacuuming, or absorbing it. But it would be harder for nature to take care of the problem.

When people are in a hurry to clean up an oil slick, certain methods are better than others. Mechanical methods are the least harmful to the environment. These methods include using booms, skimmers, and absorbent materials. Certain methods of cleaning an oil spill can be more harmful than leaving it alone. Some of these methods include making the oil sink. This just moves the problem to the bottom, where the marine life will be harmed. The surface might look better, but the damage can be much worse.

Another method that has good and bad points is scrubbing rocky coastlines with hot water. This may restore the beauty of the area quickly. But often it kills any surviving organisms that remain along the shoreline.

The short-term effects of an oil spill are terrible for the plants and animals in the area. But scientists argue about the seriousness of the long-term effects. Some say an oil spill has never been known to make a noticeable impact on the world population of any given species. Most species, these scientists claim, recover quickly even in the area of a spill. These scientists also claim that, in most cases, the environment completely recovers in less than ten years.

Other scientists have different opinions. They say that the effects of an oil spill do permanent damage, especially in sensitive areas such as coral reefs. They also say that since we can't predict long-term effects, we shouldn't take chances. No spill has ever occurred in the icy Antarctic. We know that a spill there would be harder to clean up. Because the area is so remote, it would be difficult for workers and equipment to get there. Also, bacteria would break up the spill much more slowly in water that cold. Some oil could even freeze and become permanent pack ice. The effects of such a disaster are difficult to predict. We do know that enormous amounts of krill grow in the Antarctic. One swarm occupied more than a square mile of water and went down 600 feet below the surface. Its weight was probably about one-seventh of the entire 1981 catch of all fish species all over the world. An oil spill near such an important link in the world's food chain could have devastating effects.

Of course, it would be in our best interests to avoid any future spills. But the only way we can avoid spills entirely is to stop shipping oil. It is obvious that this won't happen in the near future. So the next-best solution is to try to lessen the chances for oil spills.

Reducing the Chances for Oil Spills

On August 18, 1990, then-President George Bush signed the Oil Pollution Act (OPA). This act had three important points. The first was that ships and ports would need better navigational (steering) equipment. The second was that all oil tankers would need double hulls by the year 2010. The third was that better cleanup methods would have to be developed.

The double-hull requirement is not foolproof. An iceberg that can cut through one hull can probably cut through two. Some people argue that if a double-hull tanker runs aground, fumes might escape into the space between the hulls. These fumes could cause the tanker to explode. Another possibility is that water might gather in the space between the hulls. This would cause the ship to sink lower and leak more oil. Even worse than that, it could cause the ship to sink entirely. Despite these possible problems, several major oil companies are building double-hulled ships right now.

Until we stop burning oil, oil tankers will remain on the seas. They will get caught in storms, they will run aground, and they will leak. Perhaps our third-best hope, after reducing the chances for spills, is to find better ways to clean up the messes made by oil spills.

5
Dumping of Wastes in the Oceans

Tom's greatest joy in life was surfing. Every weekend, he was at the beach as soon as the sun came up. One Saturday morning, after eight hours of sleep, Tom was still tired. Not thinking much about it, he loaded his board into the back of his pickup truck and headed out to the beach. It was a great morning—the weather was beautiful and the waves were perfect. But Tom was so tired that he had to rest on the beach several times. At about noon, Tom and his friend Lonnie took a lunch break, as usual, at the nearby deli. But Tom wasn't *too hungry. In fact, he hadn't had much of an appetite for about a week. As Tom sipped on a soda, Lonnie noticed something unusual. "Hey, Tom," he said, "what's wrong with your skin? Your tan is turning yellow!" Tom compared his arm to Lonnie's and agreed. His skin color was definitely changing. When Tom said he was really too tired to surf for the rest of the afternoon, Lonnie knew that something was wrong. This wasn't at all like Tom, who would usually surf from dawn to dusk. Lonnie suggested that Tom see a doctor, and Tom decided to take Lonnie's advice. A few days later, he found out that he had hepatitis A. The doctor told him that it was caused by a virus sometimes found in human waste. Tom remembered reading in the newspaper that a sewage pipeline in San Diego, near Point Loma, needed some repairs. Almost 200 million gallons of partially treated sewage had been pumped into the Pacific each day for months, right near Tom's favorite surfing spot. Tom wondered if there was a connection between the sewage and his illness.*

The oceans are the final trash dump for most human waste. More than 80 percent of the pollution in the ocean comes from human activities on land. It gets there in the form of natural **runoff** from the land or in the form of trash deliberately dumped. The pollution might be plastic trash, sewage sludge, or industrial chemicals. It might be food wastes, heavy metals, or pesticides. It might even be radioactive wastes created by nuclear power plants.

Most people don't really think about what happens to their trash after it is picked up by trash collectors. They don't stop to think about what happens to the water that goes down their drains. And they don't worry about what happens to the oil that leaks from their cars.

Most of the water used in homes in the United States goes to sewage treatment plants to be purified. This includes water from the washing machine, shower, kitchen sink, and toilet. It also includes any other liquid washed down a drain. Suppose you paint your bedroom and then rinse the brushes in the sink. That paint and paint thinner go to the sewage treatment plant, too. From the sewage treatment plant, these liquids eventually go to the ocean. Some coastal cities even pump raw or partially treated sewage directly into the ocean.

Sewage Sludge

Sewage treatment plants are good at removing dirt, food waste, and some other pollutants from the water. In addition, they treat the water to kill harmful bacteria and viruses. But certain chemical pollutants cannot be treated at sewage treatment plants. Chemicals that are found in paint thinners go right through the treatment plant without being changed. Sometimes, certain **phosphates** (cleaning agents) used in detergents also pass through unchanged. When the treated water is dumped into the waterways, it eventually gets to the oceans. The untreated chemicals aren't good for marine life.

Sometimes untreated sewage gets into the oceans. In most areas of the United States, water with human waste passes through a treatment plant before being released into the waterways. Sometimes, however, a heavy storm may cause waste water to back up before it even gets to the plant. Then it might overflow into surface water without ever being treated. When this happens, disease-causing organisms get into the water. They can spread dysentery, hepatitis A, and other diseases. Hepatitis A is caused by a virus sometimes present in human wastes. The disease affects the liver. A person who has this disease suffers from loss of appetite and fatigue. Another symptom of the disease is jaundice, a condition that makes the skin look yellow. In recent years, the number of cases of hepatitis A has been rising.

Other viral diseases can result from swimming in or drinking contaminated water. For example, in one case in the summer of 1992, a surfer in Malibu, California, caught a type of virus that can cause heart disease. It is called coxsaki-B virus. The surfer had to have a heart transplant. Tests of the water off the Malibu coast showed that the virus was in the water. Although officials have not linked the surfer's illness to the water, there does appear to be a connection.

Human waste that is not treated can do other damage besides spread disease. It also robs water of oxygen. Such materials as animal waste and food waste also use up the oxygen in water.

The reason for this is that certain organisms in the water break down these materials. To do their work, these organisms must use oxygen. They might use so much oxygen that the fish and other life in the ocean can't survive.

How Coastal Areas Are Polluted

Cities are the source of most of the pollution of coastal areas. Factories, homes, and office buildings cause a great deal of hardship for the environment. But it is not just the cities that create the problems. Farms also contribute their share of pollutants to the oceans.

Industry   Factories belch out poisons in the form of nitrogen oxides and heavy metals. These have a serious effect on shellfish beds in the coastal areas. They also kill **spawning** fish (fish that are producing or depositing eggs). In addition, these poisons gather in the bodies of the fish that manage to live. When people eat these fish, they can get very ill.

One example is the mercury poisoning experienced by the people of Minamata, Japan, in the early 1950s. The illness is now called Minamata disease because this was the first place where it was recognized. Many people in this little fishing village became extremely ill after eating the fish from their area. The fish had been poisoned by mercury, and the people suffered severe nerve disorders. Hundreds of people died, and thousands of others suffered permanent brain damage. The mercury had been dumped into Minamata Bay by the

Chisso Corporation. This company was producing chemicals used in the making of plastics. Fish and shellfish in the bay were affected. As people ate them, the poisons built up in their own bodies. By 1975, there were 3,500 known victims of this disease.

Agriculture   Runoff from farms is very harmful to life in the oceans. The pesticides (poisons that kill insects and other pests) and herbicides (poisons that kill weeds and other plants) do not break down. After they get into the ocean waters, they pass through various marine food chains. At each link in the food chain, the poisons become more concentrated. That is, one animal might get some of the poison in its body. Then, that animal is eaten by another. The second animal digests the body of the first, but the poison remains in the second animal's body. The second animal might build up quite a bit of poison before it is eaten by another animal.

Another harmful runoff from farms is fertilizer. Fertilizer often has nitrogen and phosphorus in it. Plants need these chemicals to help them grow. When there is too much nitrogen or phosphorus in the ocean, however, it can upset the balance of the area. These nutrients lead to excessive growths of algae. The result is what is called an *algal bloom.* These dense blooms are so thick that they block the sunlight. Lack of sunlight kills other ocean plants below the surface. Then, as the algae die, they sink to the ocean floor. There they are broken down by bacteria, which use oxygen in the process. Because there are so much dead algae and other plants at this point, a great deal of oxygen is used. This deprives marine animals of the oxygen they need. Fish, shellfish, and other creatures will then either die or leave the area.

Some of these algal blooms produce toxins that are harmful to marine animals. For example, there was a bloom in the Atlantic Ocean off the coast of South Carolina in 1988. It was linked to about 3,000 dolphin deaths, who died after eating fish that were contaminated by the algal toxins.

Nuclear Power Plants

In power plants that produce nuclear power, there is a great deal of waste material. This material is radioactive and, therefore, very dangerous. Such waste is extremely harmful to the environment and to all living things that come into contact with it. Another danger comes from the water that is used to cool the power plant. This water is not radioactive, but it becomes very hot when it passes near the nuclear reactor. This waste water causes thermal (heat) pollution. Such pollution can kill plant and animal life that needs cooler temperatures to survive.

What to do with the waste has become a serious problem. Sometimes it is buried in containers in remote areas such as deserts. Other times it is buried at sea in strong containers. The biggest problem is that such wastes take a very long time to break down. It will take thousands of years for the danger to lessen. Some types of radioactive waste will still be deadly after 50,000 years. One source of nuclear power, uranium 238, takes 4.5 million years to lose only half of its radioactivity. Many of the containers in which this waste is stored have begun to leak. One example is the containers stored in the ocean by the Farallon Islands, near San Francisco. But nuclear waste stored in other places besides the ocean can eventually find its way to the ocean. Nuclear accidents, such as the 1986 explosion at the Chernobyl Nuclear Plant in the Ukraine, put radioactive material into the air and the ground. Rain and other elements of the water cycle will eventually take this material to the oceans.

Pablo took a little stroll along the marine preserve near his home. All along the shore were ducks, egrets, and geese. As Pablo walked along, he noticed one duck with a six-pack ring wrapped around its head. The poor animal was doing everything it could to get rid of it. It dragged its head along the beach in hopes of loosening the ring. It dove under water in hopes that the ring would slide off. "I'll save you!" said Pablo. He then tried to catch the animal when it was on land. Whenever he got close, the duck would fly away, into the water. Pablo had an idea. He would go home and get his fishing rod. Maybe he could snag the six-pack ring with a weight tied to the end of the fishing line, pull the duck in, and save it. Twenty minutes later, Pablo cast his line out toward the duck, but he missed by several yards. After trying for half an hour, Pablo noticed that the duck was getting very tired. It came in from the water and put its head down on the ground, seeming to give up the struggle. Pablo managed to sneak up and grab the duck. The duck was too tired even to run away. Up close, Pablo could see that the six-pack ring was wrapped tightly around the duck's head three times. He used his pocket knife to cut it. Finally, the duck was free. "Name your babies after me," said Pablo, as he put the happily quacking duck back into the water.

Trash from Ships and Beaches

Ships dump about six million tons of trash into the oceans each year. This trash includes glass bottles, tin cans, plastic containers, wire, wood, and food. Some wastes that are dumped into the ocean are broken down easily. Others sink to the bottom, possibly affecting marine life there. Still other types of waste are lightweight. These wastes float on the surface, tempting birds to sample them.

Not all of the plastic found in the oceans was dumped from ships. A lot of it started as trash on the beach. People leave picnic debris behind, and it blows into the ocean. This wouldn't be such a terrible problem if the plastic would decompose, but it doesn't. Scientists have found plastic items in the ocean that are 50 years old.

Many animals are interested in floating objects because they appear to be food. Sea turtles see floating plastic bags that look like jellyfish. They eat these bags and choke on them. Other animals eat pieces of plastic that fill up their stomachs, preventing them from digesting any real food. These animals may starve to death.

According to a United States government estimate, 26,000 tons of plastic packaging and 150,000 tons of fishing gear are either dumped or lost at sea each year. Fishermen have reported seeing thousands of northern fur seals wrapped in fishnet.

On beaches in Los Angeles County alone, people leave behind 75 tons of trash each week in the summer. Most of this trash is plastic. These plastics are more dangerous to marine birds and mammals than any other trash.

A company in Toronto makes a plastic that *does* break down. One of the products made by this company turns plastic to dust after it has been exposed to sunlight for 60 days. Plastics made with this product cost 5 to 10 percent more to manufacture. Several large plastics makers in the United States now make plastic six-pack rings that break down in the environment. This is a step in the right direction, but more steps need to be taken.

6

Activities of Greenpeace and Other Organizations

While scuba diving off the coast of New Zealand, Travis was surprised to see a boat on the ocean bottom. A closer look showed the ship's name, *Rainbow Warrior*, toward the front of the boat. Travis then saw the word Greenpeace in large letters along the side. Hoping to find something of value aboard, Travis swam inside the boat. He found nothing worth taking to the surface. When he told his friend Elsie what he had seen, she asked him, "Don't you know what the Rainbow Warrior is?" Then Elsie told him the story

*about the famous vessel. In 1985, the ship had
a big role in an anti-nuclear campaign. The crew
was planning to observe and report on the effects
of French nuclear testing on an island in Polynesia.
To stop them, the French secret service put two
bombs on board the Rainbow Warrior. Laurent
Fabius, then Prime Minister of France, later
admitted that the French government was responsible for these bombs. When the bombs went off
in the middle of the night, all the Greenpeace
activists except one escaped. The man who died
was a photographer named Fernando Pereira. The
father of two young children, he drowned while
trying to save his equipment. The following year,
the Rainbow Warrior was given a burial at sea off
the coast of New Zealand. In his dive, Travis had
seen not only a wreck of a boat but an underwater
memorial.*

Greenpeace is an organization that began in 1971. At first, its
purpose was to protest nuclear testing on Amchitka Island in
the Aleutians. The people who joined were concerned about
the threat of earthquakes, tidal waves, and nuclear fallout.
They were also concerned that the Alaskan island would be
completely destroyed. These were among the first people to
use public protest in the name of ecology.

Their first campaign didn't exactly do what it set out to do. The
group of 12 people set out from Vancouver in a remodeled

fishing boat named the *Phyllis Cormack*. They thought that if they simply set anchor in the blast zone, then-President Richard Nixon would call off the test. But after six weeks of being battered by strong winds and 33-foot waves, they returned to Vancouver. They hadn't even been able to get near the test site. But they did manage to attract plenty of attention from the media. The atomic test was eventually cancelled. Public opinion soon caused Amchitka to be named a bird sanctuary. The Greenpeace activists went on to become the world's most famous environmental lobbying group.

Greenpeace now has over 5 million supporters. The organization is involved in campaigns to save dolphins. It is also trying to put an end to the use of chlorine by all industries. Greenpeace was recently successful in its campaign to stop mining in Antarctica. Because of Greenpeace's efforts, 23 members of the United Nations signed an agreement. They agreed to ban all mineral and oil mining in Antarctica for the next 50 years.

Campaigning to Save the Whales

Greenpeace is perhaps most famous for its campaigns against whaling. This campaign began in the early 1970s, with the efforts of Paul Spong. Spong had worked with killer whales at the Vancouver Aquarium. He was convinced that humans could communicate with whales through music because he had seen whales respond to music. Not only that, but whales also create their own music and use it to communicate with one another. Spong decided to try to convince people in general, and Japan in particular, that whaling was wrong. Spong believed that these beautiful, intelligent, and endangered creatures needed protection.

Whale music is made up of a variety of different kinds of sounds. Sometimes whales make clicking noises, which are repeated in different combinations. Sometimes they make a loud noise, which can sound like an explosion or a shot from a gun. But the most beautiful sounds they make are very strange, and they are hard to describe. These noises can sound like wind blowing through trees. Sometimes they sound like moans and cries. In 1967, a man named Roger Payne recorded the music of the humpback whales. This became a best-selling record. The record's popularity helped to educate people about whales, and it made them more concerned about saving whales from extinction.

In 1972, 53 members of the United Nations had signed an agreement to stop whaling activities for 10 years. By 1974, Japan had not even slowed down its whaling activities. A series of lectures given by Spong in Japan didn't convince the Japanese to change. But Japan wasn't the only country to ignore the United Nations agreement. Several others, including the Soviet Union, Norway, Korea, Iceland, and Australia, still did a lot of whaling.

One famous confrontation took place in June of 1975. Greenpeace's boat, the *Phyllis Cormack,* found the Soviet whalers 300 miles off the coast of California. Two Zodiacs, small rubber boats, were lowered into the water. The Soviet ship was 750 feet long and as tall as a 10-story building. Each Zodiac was 14 feet long and used a 50-horsepower outboard motor. The contrast was like that between David and Goliath of Biblical times. The Greenpeace crew finally managed to accomplish its goal: to put a Zodiac between a harpoon and a whale. The idea was that no harpooner would take the chance of hurting a human being during a whale hunt. But the Greenpeace thinking was wrong—the harpooner fired anyway. The 250-pound cast-iron harpoon landed in the body of a whale only 10 yards from the Zodiac. As another whale was killed while charging the Soviet harpoon boat, Greenpeace cameras were still rolling. The film was shown on TV by Walter Cronkite during his evening news show. Suddenly, Greenpeace activists were heroes, and the anti-whaling campaign had gained supporters all over the world.

The fight to save the whales from extinction is far from over. Several countries continue to ignore United Nations guidelines. Japan, for example, goes on killing large numbers of whales for "scientific research." Siberian fur farmers continue to feed whale meat to caged minks. And Greenpeace activists continue to buzz around ships in Zodiacs.

But many scientists believe that it is not whaling that is the biggest threat to whales. They believe that the biggest threat to whales is the pollution of the oceans. Whale carcasses full of toxic chemicals are frequently found on beaches. Such pollution is also a target of Greenpeace activities.

Campaigning Against Chlorine

Because chlorine is a harmful chemical element that pollutes the environment, Greenpeace wants manufacturers to stop using it. This chemical is used to bleach paper, among other things. In 1991, Greenpeace worked with a German paper mill to find a different way to make paper. Then it delivered a four-ton roll of the glossy, magazine-quality paper to a Canadian magazine publisher. Greenpeace was trying to embarrass the company into using a product that was less polluting to the environment.

Campaigning Against Nuclear Testing

Greenpeace has also campaigned against nuclear testing. In 1954, long before Greenpeace even existed, nuclear testing in the Marshall Islands exposed hundreds of people to radiation. People on the island of Rongelap were exposed to about half of what was known to be a lethal dose of radiation. Even though the authorities knew the danger, the people were not moved from the island until three days after the fallout began. According to Greenpeace, 70 percent of the survivors now have thyroid cancer, burn scars, and signs of premature aging. A large number of the children of the victims are deformed and/or mentally retarded. The land and the ocean were also affected. When the islanders were returned to their homes, government officials advised them not to eat locally grown food or locally caught fish. In 1985, in answer to requests from the islanders, Greenpeace moved them to a different island.

Greenpeace continued to demonstrate against nuclear testing. Greenpeace boats have been rammed by the United States Navy, by the French Navy, and by the United States Coast Guard. Greenpeace crew members have been beaten, arrested, and jailed. The organization was not able to stop the French from testing nuclear weapons in 1985. But by 1986, 13 nations signed a treaty supported by Greenpeace. This treaty banned testing nuclear weapons and dumping radioactive waste in the South Pacific.

Greenpeace Today

Greenpeace follows certain principles that have not changed since its beginning. Greenpeace demonstrators always act in nonviolent ways. Greenpeace does not identify itself with any political party. And Greenpeace does not accept any money from corporations or governments.

Critics of Greenpeace say that the group has become too much of a corporation itself. It has a large and complex structure, with offices in 27 countries. The organization hires top researchers and lawyers. Representatives of Greenpeace discuss issues with powerful corporate and political leaders. It plays a big role in environmental debates at the United Nations. And it uses modern communications techniques to raise money and spread its message. To some people, this means that it has lost touch with its simple roots. To others, it simply means that the group is using powerful techniques to win important campaigns.

The Sea Shepherd Society

The Sea Shepherd Society has goals similar to those of Greenpeace. Its methods, however, are different from those of Greenpeace. This group was founded in 1977 by Paul Watson, a former activist for Greenpeace. Watson had been kicked out of Greenpeace because of his belief in the value of violent methods. Greenpeace, which follows the principles of Mahatma Gandhi and Martin Luther King, Jr., believes in nonviolence, so Watson was no longer welcome. Watson responded by forming his own group.

Almost single-handedly, Watson and his supporters managed to put an end to illegal whaling in the Atlantic Ocean. He had discovered that some whalers were killing protected whales and undersize whales. Some were whaling in protected waters. Some were whaling for a longer time than they were supposed to. Watson decided to put a stop to it. He strengthened the bow of his ship, the *Sea Shepherd,* with 18 tons of concrete. Then he rammed a whaling ship named the *Sierra* with his own ship. The *Sierra* suffered major damage and was eventually scrapped. Watson was arrested and later released. By the time he got his boat back, it had been stripped of anything valuable. So he decided to scrap the *Sea Shepherd.* He continued his activities later with the *Sea Shepherd II.* In addition to protecting whales, the *Sea Shepherd* Society tries to protect seals. One way they do this is to travel to the Arctic to paint seals' coats. This doesn't hurt the seals. It does, however, make their fur worthless to people who would kill them for their coats.

The Cousteau Society

Another important group concerned with the oceans is the Cousteau Society. Founded in 1974, the organization promised to be a nonprofit, educational group dedicated to saving the environment. Jacques-Yves Cousteau's films and books continue to educate people about the seas. No single person has done as much as Jacques Cousteau to educate people about the ocean and the creatures in it. But the Cousteau Society is not really involved in solving environmental problems. Cousteau has no interest in politics, and he avoids participating in political activities. His real work is in exploration, and educating young people.

Cousteau wrote a "Bill of Rights for Future Generations." In this, he declared that each generation has "a right to an uncontaminated and undamaged earth." One way to assure this, Cousteau said, was to develop new sources of energy. In 1978, he said that the United States needed to spend one trillion dollars over the next 15 years on this project. Cousteau said we should try to harness the energy of the sun, wind, moving water, ocean, and plants.

7

What Can You Do?

"**P**aper or plastic?" asked the grocery bagger when Melinda got to the checkout stand. Without giving her answer a second thought, Melinda said, "Plastic." As she carried out her plastic bags, Melinda walked by a display near the door. The store was selling cloth grocery bags." What a waste of money!" she thought. Even if I bought a few of those bags, I'd probably forget to take them into the store with me. These plastic bags are just fine."

Within a half hour, she had arrived home, unpacked the food, and put the plastic bags in the trash. Then she noticed that her trash can was full. "Not again," she sighed. "I thought I emptied that trash can yesterday." So she took her trash down to the larger cans outside. Then, she decided to take a little walk.

Melinda's apartment wasn't far from the beach. Soon, she was walking along the shore. She saw a group of birds pecking at something up ahead. As she approached them, she saw that they were taking little bites out of a plastic bag that had blown onto the sand. She noticed that a few other plastic bags were floating in the water. "That plastic can't be doing those birds any good," she thought. "In fact, it might be harming them. And I wonder what plastic might do to a fish that swallowed it." I wonder what I could do to cut down on the problem. I'd hate to be responsible for some animal getting sick. But I can't think of anything that would help."

Sometimes it seems that everything we do is bad for the environment. Each time we turn on a light, cook our food, or use hot water, we use energy. This increases the demand for more energy. In producing that energy, our power companies may pollute air, ground, and water. They may use up more of our natural resources than we can spare. Every time we throw something into the trash, we contribute to a serious problem. All that trash has to go somewhere. Often, toxins in that trash leak out into the ground and the air. Of course, our oceans are affected by any pollution of the ground and air. Each person, however, can make a difference in the amount of damage done each day. Here are a few ideas.

Bring Your Own Bags to the Supermarket

There is really only one good answer to the question, "Paper or plastic?" That answer is, "Neither," as you hand over your own cloth bags. You can use tote bags with handles, string bags, beach bags, briefcases, or your own pockets, depending on the number of items you've purchased. Another choice is simply to reuse the paper or plastic bags you used last time. Make this a habit, and you'll help to cut down on the ever-growing problem in our **landfills.**

Recycle Plastics

Plastics are designed to last. In fact, some plastics will last for 400 years in a landfill. As they break down, they often leak toxins into the ground. When this happens, the toxins will eventually get into the groundwater. Once these toxins are in the water cycle, they will eventually get to the oceans. Less than one percent of all plastics are being recycled today. Part of the reason for this is that some plastics are very difficult to recycle. A squeezable ketchup bottle, for example, uses six kinds of plastic. One kind gives shape to the bottle. Another kind gives strength. Another kind makes it easy to squeeze. The problem is that most recycling methods can process only one type of plastic. Find out what kinds of plastic can be recycled in your area. One way to do this is to call a recycling center or your trash collection company. Then, keep all recyclable plastic in a special container until there's enough to take to the recycling center. Reduce the amount of plastic you use by reusing bags and finding other uses for plastic containers.

Snip Six-Pack Rings

When a six-pack ring is underwater, it is invisible. Marine animals can't avoid them and often get them stuck around their necks. Pelicans and other birds often dive straight into six-pack rings as they are diving for fish. The ring gets caught in the bird's beak, preventing the bird from eating. Baby seals and sea lions get the rings caught around their bodies. As they grow, the rings become tighter, until the animal dies of suffocation. To prevent these problems, be sure to snip six-pack rings before throwing them away. Use scissors, a knife, or your bare hands to tear each circle so that an animal won't get hurt.

Buy Clean Detergents

Many detergents contain phosphates, which are chemicals containing phosphorus. Phosphates soften water and make it easier to clean clothes. But when phosphates get into lakes, streams, and oceans, they fertilize algae. The algae grow in large amounts. When they die, the bacteria that break them down use a lot of oxygen. Other living organisms in the water can't get enough oxygen, and soon they leave or die. To help prevent this, use detergents that are low in phosphates or, better yet, free of phosphates. Check labels. You'll find that most liquid detergents are phosphate-free.

Plant a Tree

Planting a tree is an easy thing to do, and it's very helpful to the environment. Trees remove carbon dioxide from the atmosphere, replacing it with oxygen. This in turn reduces the possibility of acid rains. Less acid rain means less polluted water entering our oceans and other waterways. If you live in a warm climate, trees have the extra benefit of helping reduce the need for air conditioning. Trees that shade houses and other buildings can cool the air inside by ten degrees. Less air conditioning means less energy use.

Use Safe Pesticides

Not too many years ago, pesticides seemed like a great idea. But the truth is now out: they can cause disastrous effects on the environment. Pesticides are designed to attack only certain pests, but the problem is that they do more than that. They often poison birds who happen to eat fruits with pesticides on them. They kill other forms of wildlife, too. They seep into groundwater and pollute drinking water. They are even destroying helpful organisms in the soil, such as earthworms. And they are washed into the oceans, affecting life there. Proof of this is the fact that DDT, a powerful insecticide, has even been found in the bodies of penguins in the Antarctic. The worst part is that the pests eventually develop the ability to resist the insecticide designed for them.

There are better choices. Some pesticides are safe and don't build up as poisons in the environment. There are also natural pesticides available that can control many different garden pests. One type of pesticide is made from the bark of a tree. Another kind is made from a flowering plant, the chrysanthe-mum. Some plants have an odor that keeps certain pests away. For instance, if you plant marigolds (a bright yellow or orange flower) around the edges of your garden, it will keep many types of insects away from flowers and vegetables. None of these safe insecticides has a lasting negative effect on the environment. Ask for natural pesticides at your local nursery. They can tell you what will work best in your garden.

Buy Organically Grown Produce

Organic foods are those that are grown without using synthetic (not natural) pesticides. Recycled organic waste (compost) is used as fertilizer. No artificial preservatives or coloring are added to organic food. Whenever possible, buy organically grown produce. If none is available, talk to the manager of the store where you shop. Urge the manager to make organic fruits and vegetables available. Also, urge her or him to buy locally grown produce. Because such food doesn't have to survive long trips to get to the store, it usually has fewer chemicals.

Avoid Pouring Harmful Chemicals Down Drains

Harmful chemicals, such as paint, paint thinner, and motor oil, pollute any water they enter. Such chemicals, when poured down drains, eventually get to the oceans. The thing to do with these items is to store them in cans with tight lids. Keep the cans in a place where small children and pets can't get into them. Save them until your area has a special garbage pick-up called "toxic waste collection."

If your area doesn't have such a service, call your community's public works department. Find out the location of the nearest hazardous waste dump. Whatever you do, don't put these cans in the regular garbage. They can do severe damage to the environment.

You might want to organize a neighborhood toxic "sweep." Most likely, no single household will have enough toxic waste to justify taking a trip to the dump. But it might be worth it for neighbors to take turns collecting everyone's toxic waste. This could be done every six months or every year, depending on the size of the neighborhood and the amount of toxic waste produced.

There are other items besides motor oil, paint, and paint thinner that shouldn't go in the regular garbage. These include car batteries, oven and drain cleaners, mothballs, floor and furniture polishes. Also included are brake and transmission fluids, antifreeze, rug and upholstery cleaners, pesticides, herbicides, and furniture strippers.

Adopt a Beach

Get together with some friends and clean up part of the beach. Fill cardboard boxes with any litter that might be harmful to sea creatures. You'll probably find plastic items left over from beach picnics. You might find fishing line, fishing nets, or kite string. Sea creatures can become tangled up in these things. You might find balloons and six-pack rings, which are very dangerous to sea animals. Make a game of collecting litter. See who can find the most unusual items. If you find aluminum cans or glass bottles, take them to a recycling center.

Write Letters to Government Officials

Write letters to officials at local, national, and international levels. Tell them your opinion about various environmental issues. Often, these officials will count the number of letters they get. If they hear from many people about the same problem or issue, they are more likely to do something about it. Make your voice heard—you *can* make a difference!

Glossary

ambergris A substance found in the stomachs of some whales. It is used to make some perfumes.

aquatic Growing in, living in, or having to do with water.

biodegrade To break down into harmless parts, through the action of living things.

bioremediation The use of living organisms to clean up an area damaged by an oil spill.

bycatch Fish that are caught but not wanted. They are thrown back into the water.

carbon dioxide An odorless, colorless gas breathed out by animals and used by plants to make food.

coral reef A ridge made of coral near the surface of the water.

drift Deep sea current.

drift net A huge fishing net made of thin, strong nylon. It drifts in the water, trapping any animal that comes by.

ecosystem A community of plants and animals, and their environment, acting as a unit in nature.

environment Conditions and surroundings that affect the development of living things.

estuary A place where rivers and oceans meet.

evaporate To change from a liquid into a vapor or a gas.

finfish Fish that have fins rather than shells.

gill net Similar to but smaller than a drift net, it is used in waters close to the coast.

gyre A giant loop in a surface ocean current.

krill A tiny sea animal (a type of zooplankton) that looks something like shrimp.

landfill An area where trash and garbage are buried in layers of soil.

long-lining A method of catching fish. It uses long lines with scores of hooks, stretched out across the sea.

mangrove A tidal wetland in a tropical zone.

marine biologist A scientist who studies ocean life.

migrate To move from one place to another to live.

nutrient Something that gives nourishment, such as food.

ocean front   An area where cold and warm ocean waters meet.

organism   An animal or plant.

overfishing   Catching so many fish of one species that they can't reproduce fast enough for the next year's catch.

papyrus   A tall grass that grows near rivers.

pesticide   A poison used to control pests, such as insects.

phosphates   Compounds of phosphorus and oxygen that are used in fertilizers and as cleaning agents.

phytoplankton   Single living plant cells in the ocean.

plankton   Tiny ocean plants that provide food for millions of ocean animals.

polar regions   Areas near the North and South poles.

pollution   Dirtiness in the air, soil, or water that is harmful to the environment.

polyp   A tiny ocean animal that lives with millions of others in a colony. The calcium from their bodies forms coral.

purse-seine net   A huge net that is placed around a school of fish and then closed like a purse.

reef   A ridge made of coral, sand, or rocks near the surface of the water.

roe   The eggs of fish.

runoff   The part of precipitation on land that eventually washes into waterways.

salinity   The amount of salt in the ocean.

saltmarsh   A tidal wetland in a temperate zone.

shellfish   Fish that have shells rather than fins.

shoal   A shallow place in the water, or a sandbar.

sonar   A device that sends sound waves and picks up the echoes with a microphone to locate underwater objects.

spawning   Producing or depositing eggs, as in the case of fish and some other aquatic animals.

toxin   A poison.

tropics   Areas of the world near the equator.

upwelling   A place in the ocean where cold water rises, bringing nutrients to the warmer upper layers.

zooplankton   Tiny animals that live in the ocean.